Yesterday & Today

AGRICULTURE

Dianne Hansen

BLACKBIRCH PRESS
An imprint of Thomson Gale, a part of The Thomson Corporation

THOMSON
GALE

Detroit • New York • San Francisco • San Diego • New Haven, Conn. • Waterville, Maine • London • Munich

THOMSON
★
GALE

© 2005 Thomson Gale, a part of The Thomson Corporation.

Thomson and Star Logo are trademarks and Gale and Blackbirch Press are registered trademarks used herein under license.

For more information, contact
The Gale Group, Inc.
27500 Drake Rd.
Farmington Hills, MI 48331-3535
Or you can visit our Internet site at http://www.gale.com

ALL RIGHTS RESERVED
No part of this work covered by the copyright hereon may be reproduced or used in any form or by any means—graphic, electronic, or mechanical, including photocopying, recording, taping, Web distribution or information storage retrieval systems—without the written permission of the publisher.

Every effort has been made to trace the owners of copyrighted material.

Picture Credits: Cover: © Archivo Iconografico, S.A./CORBIS (top), Photos.com (bottom)
© Bettmann/CORBIS, 19; © Blackbirch Press, 12
© Uthaiwan Boonloy/EPA/Landov, 29; Corel, 4
© Werner Forman/Art Resource, NY, 5
© Lowell Georgia/CORBIS, 23, 28
© Getty Images, 20
© Giraudon/Art Resource, NY, 11
© HIP/Art Resource, NY, 6 (below)
© Hulton/Archive by Getty Images, 14, 17, 18 (below)
© Bildarchiv Preussischer Kulturbesitz/Art Resource, NY, 8
© Erich Lessing/Art Resource, NY, 6 (above), 7, 10
Library of Congress, 12, 22, 24, 25, 26
© Museum of the City of New York/CORBIS, 13
© North Carolina Museum of Art/CORBIS, 16
Photos.com, 14, 15, 18 (inset)
© Jim Richardson/CORBIS, 30
© Joseph Sohm; ChromoSohm Inc./CORBIS, 26-27
© Charles E. Steinheimer/Time Life Pictures/Getty Images, 21
© Emile Wamsteker/Bloomberg News/Landov, 31

LIBRARY OF CONGRESS CATALOGING-IN-PUBLICATION DATA

Hansen, Dianne, 1946-
 Agriculture / Dianne Hansen.
  p. cm. -- (Yesterday and today)
 Includes bibliographical references and index.
 ISBN 1-56711-827-5 (hard cover : alk. paper)
 1. Agriculture--Juvenile literature. I. Title. II. Series.
 S519.H3785 2005
 630--dc22

Printed in the United States

Table of Contents

The First Farmers	4
Early Farm Villages	6
Water Management	8
Crop Rotation	10
Farm Machinery	12
Crossbreeding	14
Fertilizer	16
Pest Control	18
Preserving Food	20
Gas-Powered Tractors	22
Electrifying the Family Farm	24
Soil Conservation	26
Genetic Engineering	28
Agriculture Today	30
Glossary	32
For More Information	32
Index	32

A prehistoric rock painting from Utah shows a hunter making a kill. Early humans were hunter-gatherers.

The First Farmers

The earliest humans were hunter-gatherers. They hunted herds of wild animals. The meat was nutritious food. In addition, the people used animals' hides, antlers, and bones to make warm clothing and useful tools. As they followed the herds, people gathered nuts, fruits, and berries to eat. They also gathered wild wheat and ground it between stones to make flour.

Humans began to settle in an area known as the Fertile Crescent. Israel and Jordan are the western portion of this crescent, which also includes Iraq and the mountains of Iran. Its ancient name was Mesopotamia. The combination of excellent climate, soil, and rivers in this region produced natural fields of wheat and barley. It was an ideal place to settle.

Once settled, people began domesticating, or taming, animlas. Domesticated wolves became the first dogs. They were very good hunting helpers, and warned people when dangerous animals approached. Dogs were also trained as herders. They helped humans round up wild sheep and goats so that sources of food would always be nearby. Archaeologists, scientists who study ancient humans, have found evidence of animal herding as far back as eleven thousand years ago, in the area that is now Israel.

Farming also began around this time. When people realized that new plants would grow from seeds they collected, they saved some and planted them. These people were the first farmers. They learned that some seeds grew healthier plants than others and saved the best kinds. It was difficult to save the seeds. Wet seeds got moldy or sprouted too early. The first farmers poked holes for the seeds in the soil with a stick. If there was enough rain, the plants grew. The farmers harvested the grain by shaking the seeds into folded animal skins. If there was enough grain to last until spring, they decided to stay where their plantings were successful. When a settlement became permanent, its population began to grow.

Tools from Bones

Early farmers sometimes used animal bones to harvest wheat and barley. They bent the plant stalks over folds in their clothing. Then they beat the seed heads with the bone until the seeds were released.

This Egyptian stone block shows a dog chasing goats outside a village. Dogs helped early humans with hunting and herding.

Prehistory
500 B.C.
100 B.C.
A.D. 100
200
500
1000
1200
1300
1400
1500
1600
1700
1800
1900
2000
2100

Early Farm Villages

People gathered in groups where there was enough water for themselves and their plants and animals. Family groups gradually formed small villages. The large number of people in a village offered safety from predators. In addition, when all the grain fields ripened at once, there were many people to help with the harvest.

The grain crops by themselves did not provide enough nutrition. Meat was still a large part of the diet. Farmers could not go hunting for meat when they were tending fields, however. Growing families began to breed wild pigs and cattle so that they could always have meat. Then, even more grain was needed in order to feed the livestock.

The idea of farming began to spread from Mesopotamia. Farming villages formed in nearby Greece. Villages also grew in India, where the people planted wheat and barley. Rice was grown in southern China. The plants that grew wild in each location were the first ones to be domesticated.

FAST FACT

Tamed pigs proved to be very helpful to people who lived in villages. In addition to providing meat, the pigs kept the villages cleaner by eating the scraps that humans did not want.

Egyptian men use sickles to cut wheat in this ancient wall painting. A prehistoric sickle with a sharpened stone blade is shown at left.

As farming moved farther away from rivers, irrigation channels like this one were dug to provide fields with water.

Early farmers invented tools to suit the kinds of crops they grew. The Europeans created simple plows from forked tree branches. They dragged the branches through the soil to cut shallow trenches, called furrows, into which they spread their seeds. To dig water channels for rice plants, the Chinese made stone shovels with wooden handles. Harvesting knives, called sickles, were made with blades of sharpened stone. Harvested grains were kept dry in bins made of sun-dried clay.

As fields got larger, they grew farther away from the rivers that provided water. Mesopotamian farmers began to dig irrigation channels to let water flow through their fields. In New Guinea, farmers faced the opposite problem. Their fields were too wet, so they dug drainage channels to lead the water away. In this way, people began to make the first permanent changes to the land.

FAST FACT

The Chinese drilled holes into flat stones to make shovels. They pushed strips of leather through the holes and tied the stones tightly to a wooden handle.

A man in ancient Egypt waters crops with a shadoof, an efficient device for moving water.

Water Management

As farm families grew, small villages gradually became cities. Cities full of people required even more food. Fields had to be extended into new areas to produce more crops. Farmers looked for ways to get the proper amount of water to the crops they raised.

In China, people worked together to level fields so they could be submerged (covered beneath a shallow layer of water) at a depth best for growing rice. European farmers who lived in wet climates developed drainage systems to remove spring floodwater, which damaged their fields by carrying away fertile topsoil. In the dry desert climate of Egypt, however, villagers on the Nile River counted on the spring flood each year. The overflow covered their flat, sandy floodplains with silt, rich black soil carried down by the river from higher land. These villagers built elaborate irrigation systems to bring water through the flat fields.

People created methods to move water more efficiently and to reduce the amount needed in distant fields. The

Archimedes' Screw

Both the Mayans and the Greeks separately invented a water-moving device that the Greeks called Archimedes' screw. This was a hollow cylinder coiled in the shape of a screw, which was set into water. When the screw was turned, it lifted water at the bottom of the cylinder to the top. Water was transported in this way from low-lying lakes to mountain fields.

Egyptians invented the shadoof, a wooden framework supporting a long arm that moved up, down, and sideways that made it easy to lift buckets of water and swing them to the irrigation channels to be emptied. The skilled Mayan farmers of South America built raised planting beds that put the irrigation channels closer to the roots of their crops. This made crops easier to irrigate, and less water was lost to evaporation.

When the Roman Empire expanded to include Egypt, Greece, Mesopotamia, and most of Europe, the Romans developed ways to bring water from rainy areas to fields in dry areas. They invented concrete, which they used to build dams to create lakes and reservoirs. Concrete was also used to make aqueducts, long channels that carried water down from the hills directly to the fields and cities.

With the ability to move and store water, farmers planted crops in new locations. Expanding populations became less dependent on local lakes and rivers.

Men in ancient Greece use an Archimedes' screw to transport water.

Crop Rotation

This mosaic shows Roman farmers plowing and sowing seeds. The Romans rotated crops to improve harvest quality and size.

Water carries dissolved nutrients to plant cells, but most of the nutrients must come from the soil. Rich soil was needed by Roman farmers, whose warm climate sometimes allowed them to plant for a second harvest. The second harvest, however, was usually smaller and weaker than the first. This was because each crop removes certain nutrients from the soil and deposits some unneeded nutrients. The solution to this problem of less successful harvests was crop rotation.

The practice of crop rotation began when Roman farmers noticed larger, healthier harvests in fields that had just grown different crops. The second crop absorbed different nutrients, including some the first crop had deposited as waste. Farmers began planning ahead to rotate crops of wheat, barley, and rye in their fields for the maximum harvest.

The wet, harsh weather in northern Europe required a different kind of farming, however. European rivers carried topsoil away and left poor soil for growing crops. To improve their soil, Europeans divided their fields into three strips. One was sown (planted) with wheat or rye. The second was sown with oats or beans. The third was left fallow, or unplanted, to allow the soil to rest and regain some nutrients.

FAST FACT

At Heathrow Airport in London, England, archaeologists have discovered that planting field boundaries made in 2000 B.C. have remained the same for four thousand years. They are still shown as field boundaries on today's maps.

Instead of rotating crops to enrich the soil, South American farmers planted beans, squash, and corn together. These crops use different nutrients, and each deposits nutrients in the soil that are needed by the others.

As colonist populations in North America grew, the colonists looked for ways to increase their harvests. They discovered that adding limestone improved the soil. European and American scientists analyzed soil to see what chemicals it contained and began to look for new sources of those chemicals.

Early European farmers divided fields into three strips. Two strips were sown with crops, while the third was unplanted.

11

Farm Machinery

Farmers in southeastern America grew cotton, an important crop for trade with Europe. In 1793 Eli Whitney invented the cotton gin, a rotating cylinder with metal bristles that pulled cotton through narrow slots, quickly separating soft cotton fiber from its sticky seeds. This made harvesting faster and easier. Farmers planted larger crops and their cotton profits grew.

Pioneers who reached the prairies in the West found rich soil, but their iron-tipped wooden plows were not strong enough to break into it. Instead, the plows broke. Pioneers were far from the stores in towns, so if they went for new parts, they would miss the planting season.

In 1797 a cast-iron plow was invented that could turn the tough American prairie sod. It was followed in 1819 by an English-made plow with interchangeable parts. Farmers could keep an extra part in case one broke. A little later, an almost unbreakable plow was developed in America from a new product called steel.

Eli Whitney invented the labor-saving cotton gin in 1793.

Whitney's invention made harvesting cotton easier and faster, and increased farmers' profits.

The harrow, a frame with teeth or discs, was used to break the freshly plowed soil into finer textures. After the crop was planted, the harrow was again drawn through the fields to cover seeds or tear up weeds.

The mechanical reaper, patented by Cyrus McCormick in 1834, cut crops with a wheel-powered blade. It sawed through stalks of grain as the wheel turned. It was followed in 1837 by a threshing machine, patented by Hiram and John Pitts. Powered by horses walking on a treadmill, the thresher knocked grain from its stalks with a revolving cylinder and blew away the chaff (straw) with a fan.

The expanded use of mechanized power greatly reduced the time required to plant and harvest corn and wheat. New factory-made steel machinery allowed farmers to cultivate bigger fields. It was also very expensive, however. Farmers had to grow additional crops to sell (known as cash crops) so that they could afford the new machinery.

Threshing Crews

Neighbors shared the expensive threshing machines and worked together to bring in the harvest. Threshing crews traveled from farm to farm with the huge machines. They spent several long days working at each farm.

A farmer uses oxen to pull a plow. The unbreakable steel plow helped farmers get through the planting season.

Luther Burbank (right) pioneered experiments with food crops and crossbreeding. Corn (below) was hybridized to create larger harvests.

Crossbreeding

Plants had been improved with specialized breeding since the earliest farmers selected seed from the healthiest wheat to plant for the new harvest. Crossbreeding began when they intentionally paired the two plants with the most wheat grains to obtain the most productive offspring.

In the nineteenth century, farmers realized they could combine, or cross, two different species or kinds of wheat to get the most desirable traits from each. This resulted in a completely new type of plant called a hybrid. Hybrids sometimes happened naturally, which increased the variety of living things. Farmers used crossbreeding to raise crops that better suited human needs.

The same crops were grown for centuries without much change, but in the early 1800s farmers began experimenting with their crops to produce plants for specific new purposes. Luther Burbank was one of the first scientists to experiment with food crops. In 1871, after hearing that diseased potatoes caused a famine in Ireland, Burbank developed a potato that would resist the potato blight disease. The new potatoes were shipped to Ireland and helped stop the famine. Burbank started experimental farms devoted to developing nutritious and disease-resistant foods.

By the 1880s, two varieties of corn were hybridized for the sole purpose of creating bigger harvests. Experiments continued for corn that was disease resistant and for varieties that could grow in many different conditions. High-yield, disease-resistant, hybrid corn seed was quickly accepted by U.S. farmers. Soon, specialized varieties suited to many soil conditions and climates were grown around the world.

Many varieties of wheat, rice, and other cereal crops were eventually grown from hybrid seed, as were vegetables such as onions and tomatoes. Varieties bred to mature at the same time were more uniform in size, which made it easier to process them by machine. Plants were cross-bred to benefit agriculture in many ways, including resistance to insects and longer storage life.

FAST FACT

More of the world's surface is planted with wheat than with any other crop. The flour made from grinding its seeds makes bread, cereal, crackers, cookies, spaghetti, and hundreds of other products.

Wheat (below) was one of many crops grown from hybrid seed, which made the crops resistant to disease and insect damage.

Fertilizer

After each harvest, nutrients that plants have removed from the soil must be replaced if the soil is to remain productive. Nutrients normally come from decayed plant and animal material and dissolved minerals. These are organic fertilizers.

Organic fertilizers have been used for centuries as farmers spread plant and animal waste on their fields. As crop production grew, however, fertilizer was needed more quickly than could be produced by the natural decaying process.

Commercial fertilizer production began in Europe in the 1840s. By 1849 mixed chemical fertilizers were sold commercially in the United States. They were so effective at improving soil that by 1910 American farmers were using approximately 3,738,300 tons of fertilizer each year.

Manufactured, or inorganic, fertilizers can be made from certain minerals or synthetic (man-made) substances to replace the most often depleted nutrients: nitrogen, phosphorus, and potassium. The nitrogen fertilizers are made mainly from ammonia gas, which is compressed until it becomes a liquid. Phosphorus fertilizer is made by grinding a mineral called apatite into a powder or by treating it with sulfuric acid to make liquid fertilizers. Potassium fertilizers are mined from deposits of potassium chloride.

In the 1920s two German chemists named Fritz Haber and Carl Bosch developed a successful method of pressurizing ammonia. Ammonia plants were built in Europe and the United States, producing an abundant supply of nitrogen. Phosphorus came first from rocks in Florida and then from North Carolina and North Africa. Potassium was first mined in

A nineteenth-century farmer spreads organic fertilizer. Commercial fertilizers produced in Europe became available to American farmers by midcentury.

Germany, but deposits were later discovered in other European countries as well as in Russia, the United States, and Canada.

The demand for fertilizers grew along with their production. American farmers purchased almost 6,845,800 tons of fertilizer in 1930, nearly twice the 1910 amount, and the trend continued. As crop production increased, surplus grain was shipped to other countries. Commercial fertilizers changed the way the world was fed.

FAST FACT
When water drains through croplands, excess fertilizer is carried to lakes and streams. It increases algae growth, which reduces the water's oxygen supply for fish and other water organisms.

Demand for fertilizers grew with farm size. These farmers sprinkle chemical fertilizers by hand over a field.

Pest Control

Eliminating weeds and insects increases crop production. Weeds compete with the crop for water and nutrients. Insects eat the crop and spread plant diseases.

Early farmers controlled weeds by digging them up with the hoe and the harrow. To control insects, they planted crops that resisted certain pests among the crops that attracted them. Birds and beneficial (helpful) insects helped by eating destructive insects. By the 1800s, soap, sulfur, and tobacco solutions were used to eliminate insects and fungus.

Large fields cultivated by machinery required long strips of a single crop, which attracted specific pests. In 1850 farmers welcomed the introduction of two effective natural insecticides, rotenone and pyrethrum, which were made from the roots and flowers of plants.

Insects damage crops and spread disease. Early farmers planted pest-resistant crops with other crops to keep insects away.

Farmers try to scare off a swarm of locusts. Insect invasions can destroy crops.

Inorganic pesticides, usually made with the poison arsenic, were developed in the 1870s. One, called Paris green, was so widely used that in 1903 the United States and Great Britain created the first laws governing the amount of pesticide that could be used. Cyanide, formaldehyde, and other poisons proved effective against certain groups of insects during the early twentieth century.

Spraying and dusting equipment was developed to distribute liquid and powdered pesticides. Carbon tetrachloride, a poison gas, was used to fumigate stored grain. In 1918 an airplane was used to dust crops for the first time.

By the 1920s poisonous pesticide residues were sometimes found in fruits and vegetables. People became alarmed. Research focused on making less dangerous pesticides. More emphasis was given to biological controls, which eliminate pests by introducing living organisms that eat them. Scientists also searched for new ways to develop pest-resistant plants.

Farmworkers spray a cotton field with pesticides in the 1920s, a time that brought awareness of pesticide dangers.

The Harmful Effects of DDT

Developed in the 1940s, the chemical DDT was found to be effective against more than five hundred pests. It was regarded as a miracle. In 1962, however, Rachel Carson's book *Silent Spring* sounded a public alarm by revealing the harmful effects of chemicals on the environment. Following its publication, many environmental policies were changed. In 1972, for instance, DDT was banned in the United States.

Prehistory

500 B.C.

100 B.C.

A.D. 100

200

500

1000

1200

1300

1400

1500

1600

1700

1800

1900

2000

2100

19

Women on a production line can green beans. Canned food remains fresh for a long time.

Preserving Food

As farmers succeeded in producing extra food, preservation and storage became the next challenge. Before refrigeration, people in cities had to shop daily for fresh foods. Vegetables and dairy products often spoiled during the lengthy wagon trips from prairie farmlands. The arrival of the railroad steam engine allowed faster food shipment.

When railroad tracks extended across North America, new vegetables from the West Coast were shipped eastward. Food was kept from spoiling on the long journey with a refrigerated railroad car. Patented in 1867 by J.B. Sutherland of Michigan, the car was insulated and vented air by gravity through ice bins and into the car.

French scientist Nicholas Appert discovered in 1810 that food heated while in sealed containers was preserved as long as the containers remained tightly sealed. This

FAST FACT

The Romans made the first major advance in storage. They invented glass bottles, which kept food fresh longer by protecting it from air and moisture.

process was used for canning food in glass jars and in Peter Durand's 1810 invention, the tin can. Made at first of tin-coated iron, cans had to be hammered open. They gained greater acceptance after 1858, when Ezra Warner of the United States patented the first can opener. Production was increased in the 1880s with the introduction of automatic can-making machinery.

Compressed gas refrigeration was in commercial use in the 1890s, and electrical home refrigerators were available by the 1920s. Truck refrigeration was not available until 1949, however, when Fred Jones invented a roof-mounted compressor.

In the 1920s cans were improved with zinc-based linings that preserved food longer. Cans for beverages, developed in the 1930s, were opened by punching holes in the top with a small triangular steel cutter. In the 1930s Clarence Birdseye began selling frozen meat and vegetables from refrigerated display cases in eastern U.S. stores. At first, Americans did not trust frozen food, but later products such as French fries, meat pies, and pizza gained quick acceptance.

The ability to preserve food was a huge development in agriculture. Cans and refrigeration made more foods available throughout the year and changed the foods people ate.

> **FAST FACT**
> In early-twentieth-century homes, iceboxes cooled butter and milk with blocks of ice that had been cut from frozen lakes.

Workers spray ice on fresh vegetables in a railroad boxcar. Refrigerated cars kept foods fresh during shipping.

Gas-Powered Tractors

In 1900 farmers relied on horses for personal transportation and delivering goods as well as for powering farm machinery. The average farm had three or four horses. Two of them were often needed only for plowing during planting season, but were maintained throughout the year. Much time and energy was devoted to the health and care of work animals, and about 20 percent of cropland was needed for animal feed. Many farmers welcomed the idea of a reliable, affordable tractor that could work without rest, did not get sick, and did not eat the crops that were planted.

Practical gasoline engines had been developed by Karl Benz in 1885 and Gottlieb Daimler in 1886. Many

Driverless Tractors

Twenty-first-century tractors may be driverless. Computers will guide tractors through fields and report on machine functions. Global positioning systems will report tractor locations, allowing farmers to attend to other business while their tractors work the fields.

Farmers welcomed Henry Ford's inexpensive Fordson tractor in 1917 because it freed them from keeping work animals.

Tractors like the Farmall improved life for farmers by lowering costs and increasing crop harvests.

inventors experimented during the 1890s with gasoline engines as a power source for farm machinery. Charles Hart and Charles Parr built the first gasoline-tractor factory in the United States and produced the first commercial tractor in 1901.

The first tractors were huge and very expensive. Hart-Parr and other early manufacturers (Case, Rumely, and International Harvester) worked to make them smaller and more affordable. In 1917 Henry Ford succeeded by manufacturing the Fordson, a small tractor offered for a much lower price. This simple machine was a 20-horsepower engine on four steel wheels that had a hitch for pulling farm equipment and an open-air metal seat located directly behind the engine.

In 1925 International Harvester produced a successful innovation, the general purpose (or GP) tractor. Called the Farmall, it was sold along with complementary machinery for cultivating, plowing, and harvesting.

Gasoline tractors steadily replaced the horse in the United States and Europe. They powered larger farm implements, processing more crops at one time. With tractors, farmers were able to reduce costs and increase food production.

23

Electrifying the Family Farm

> **Some Uses of Electricity**
>
> Reliable electric radios brought current news of the outside world to the farm, as well as entertainment that soothed worries about weather and crops. Farm business accounts were tended in the evening by electric light. Electric brooder houses kept baby chickens warm.

At the beginning of the twentieth century, most farm communities were self-sufficient (producing everything needed to survive). Farmers grew all their own food. They kept an ox or horse for a work animal, a cow for milk and butter, and chickens for eggs. Children gathered eggs, milked cows, and hauled fuel and water. They also cleaned barns, fed farm animals, and herded cattle. Care of the crops, the animals, the machinery and buildings, and the family's daily needs kept large families very busy and was tied to the seasonal schedule of the farm.

City homes and markets enjoyed the first electric refrigerators in the 1920s. Farmers, however, isolated in fields far from the new electric lines, continued to live as they always had. By 1930, 58 percent of farms in the United States had cars and 34 percent had telephones, but only 13 percent of farms had electricity. For most farmers, the workday began before dawn and ended at sunset. Light in farmhouses and barns came from dim and dangerous kerosene lamps.

Once electric power was brought to most U.S. farms in the late 1930s, farmers could safely work after dark,

Electricity on farms allowed farmers to use more efficient machinery such as this electric corn sheller.

Refrigeration machines such as this electric milk cooler helped prevent food-borne illnesses.

feeding animals and repairing machinery. Electric machines and tools were safer and more efficient.

Electric pumps brought water to the fields, improving irrigation. They also brought water directly into the farmhouse, eliminating the water-hauling chores. Along with electric stoves, running water in the kitchen made food cleaning, preparation, and preservation easier and more sanitary. Electric refrigerators safely stored fresh foods for home use and for the market. Food-borne illnesses declined. Increasingly efficient, farms grew in size and productivity.

Decline of the Family Farm

By the 1960s the number of family farms was declining. Some were forced out of business when costly crop failures prevented payments on the expensive new machinery. Others were purchased by large agribusinesses, groups of companies specializing in all the areas needed for maximum crop production.

25

The Dust Bowl disaster forced many prairie farmers to abandon their farms in the 1930s.

Soil Conservation

Power machinery made life much easier for the farmer, but intense farming was creating problems in the soil. Each year, more fields in the United States were cultivated until much of the American prairie was a monoculture (growing a single crop) producing only wheat, the most profitable crop.

American farmers failed to learn from earlier experiences with monocultures, which had resulted in

FAST FACT

Dust Bowl was a term first used by a newsman in the 1930s to describe the states hardest hit (Texas, Oklahoma, Kansas, and Colorado) by drought and dust storms.

disaster. They continued to plow the prairie until several years of record drought in the 1930s brought repeated crop failure. Many farms were abandoned. Dry fields extended for miles with no vegetation to hold the soil in place.

Driven by fifty-mile-an-hour winds, giant dust clouds that stretched a hundred miles wide picked up the topsoil and carried it away. Vast areas of the prairie were destroyed over a three-year period. The Dust Bowl disaster affected not only farmers but all the people who relied on them for food.

Scientists, farmers, and the government worked together on new methods of conserving soil and water to prevent future disasters. They analyzed soil to discover how to make it more fertile. They developed contour plowing methods, plowing circles around a hill instead of up and down. This allowed water to sink into the soil instead of running off with it. Farmers planted trees in long strips that would protect the soil from the wind and hold more water in the soil.

New methods of soil protection were developed. Cover crops were planted after the harvest to return nutrients to the soil and hold it in place. With proper soil conservation, future weather disasters could be minimized.

The Irish Potato Famine

In the nineteenth century, poor Irish farmers all grew the same variety of potatoes in the tiny plots allowed them, eating little else. Potatoes provided more nutrition than anything else that would grow in Ireland's thin soil and cool weather. When the 1845 potato blight killed the crop for several years, over 1 million Irish people starved to death.

Farmers now focus on soil conservation and contour plowing, the method used for planting these wheat fields.

Genetic Engineering

The science of genetic engineering emerged in the late twentieth century. Genetics is the study of heredity, or how organisms (living things) pass on their traits to the next generation. Each cell in an organism carries all the genes that determine such things as size and appearance. In humans, for example, body type, height, and eye color are inherited from parents through their genes. Genetic engineering takes place when genes from the cells of one organism are inserted directly into the cells of another one.

Genetic engineering is different from hybridization, in which changes are made by selecting parents until the desired trait is obtained. Hybrids are bred from members of the same family, as when related species like broccoli and cauliflower are crossed to make broccoflower. In genetic engineering, genes from an unrelated species are inserted in a way that could not occur naturally. For

A researcher cuts seeds from a sunflower for use in genetic experiments. Scientists genetically engineer seeds to create desired traits.

example, tomatoes, which are sensitive to frost, can be made more tolerant of cold temperatures with the insertion of the gene for cold tolerance from a fish.

Genetically engineered (GE) foods are sometimes called transgenic or GMO (standing for "genetically modified organisms"). Plant seeds are usually altered for two reasons. Some are inserted with genes from bacteria that kill insects. Others are made more resistant to the chemical herbicides used for weed control. As a result, more herbicides can be used without killing the food plant.

Since they contain substances never before found in food, GE food products have caused debate all over the world about the possible risks to health and the environment. Most European countries distrust and reject GE trade crops. Many southern African nations refused GE food offered during a famine.

About forty GE crops are approved for sale in the United States. These include corn, soybeans, squash, tomatoes, and potatoes. Most foods processed in the United States now contain some ingredients made from GE corn or soy. There is much public debate about whether to label products containing GMOs, so people can choose other foods if they prefer.

Protesters in Thailand demonstrate against GMO foods. Many countries have rejected these foods.

Terminator Technology

"Terminator technology" is a nickname farmers and environmentalists gave to a new GE process that prevents a crop's seeds from sprouting for a second planting. Farmers would be forced to buy new seeds each year. Poor countries, however, depend on saved seeds. India banned importation of these GE seeds.

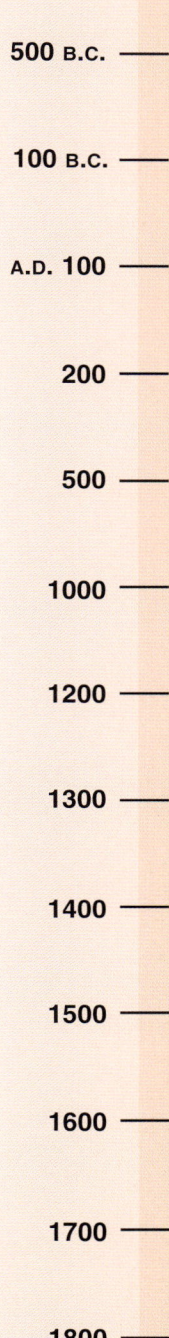

29

This food processing worker monitors eggs receiving light treatment, a new technique in food preservation.

Agriculture Today

There are many more people on the earth today than when farming began. Today's scientists look for ways to grow enough food for everyone. Many important discoveries have already been made.

Breeding experiments have produced more and bigger animals. Scientists have shown that animals can be cloned, or reproduced without parents.

Many common food plants have been genetically engineered. Some people believe this is a safe way to increase harvests and eliminate hunger in the world. Others, however, think it will harm the environment and make hungry nations dependent on agricultural companies for their food.

Much of today's food is processed. Chemicals and dyes are added to make foods last longer, look nicer, or pour more easily. Machine processing sometimes destroys nutrients and makes food less healthy. When huge quantities are processed, contaminated food is sometimes not discovered. Inspection becomes important to protect people from diseases such as E. coli.

Irradiation kills the bacteria on food, but it also creates new chemicals that do not naturally occur in the foods. Many consumers want more safety studies conducted before purchasing irradiated foods.

Many organizations work to keep both food and land healthy. Some look for ways to reduce the use of chemical fertilizers to keep water clean for the plants and animals that use it. Others try to reduce the use of chemical pesticides that end up in food.

Organic farmers produce food the way it has been done for hundreds of years, without chemical fertilizers, pesticides, and herbicides. This requires more labor, so organic foods usually cost more. Many supermarkets, however, now find it profitable to add organic foods to their shelves.

Though they seem new, the challenges of agriculture in the twenty-first century are the same ones faced by the Egyptians five thousand years ago. Humans still search for ways to grow the greatest amount of food with the least effort and for the most profit.

Organic Crops

The U.S. Department of Agriculture has certain requirements for food that will be labeled "organic." A government inspector visits the farm where the food is grown to make sure the farmer is following all the rules governing organic crops. Companies that process or package the food must also be certified (approved).

Consumers worried about chemically treated foods are willing to pay more for organically grown foods.

Glossary

agribusiness: A group of companies specializing in a variety of agricultural needs, such as machinery, food processing, chemical manufacturing, and finance.
aqueduct: A tube built for carrying water.
archaeologist: A scientist who studies the remains of past human societies.
conserve: To keep safe from waste or destruction.
contaminate: To soil, infect, or make unfit for use.
cover crop: A crop planted after the cash crop for the purpose of holding soil in place and returning nutrients to it.
crossbreed: To cross two varieties within the same species.
cultivate: To prepare to help crops grow.
drought: A long period of dryness that prevents the growth of crops.
Dust Bowl: A region that suffers from prolonged droughts and dust storms.
fallow: Left unplanted or inactive after plowing.
fertile: Capable of producing plant growth in great quantities.
fertilizer: A substance used to make the soil more fertile.
hybrid: A plant that is created by crossbreeding two similar species.
irrigate: To supply dry places with water.
monoculture: The repeated growth of a single crop over a large area.
organic: Produced naturally from plants or animals without the use of any chemicals, antibiotics, or pesticides.
sow: To scatter seed on the earth for growth.
thresh: To separate seeds from a harvested plant.
topsoil: The surface layer of soil in which plants have most of their roots.
yield: The amount of harvest produced.

For More Information

Books

Ned B. Halley, *Farm*. New York: Alfred A. Knopf, 1996.
R.J. Stephen, *Farm Machinery*. New York: Franklin Watts, 1986.
Michael Woods and Mary B. Woods, *Ancient Agriculture: From Foraging to Farming*. Minneapolis: Runestone, 2000.

Web Sites

America's Story from America's Library (www.americaslibrary.gov). Some of the fascinating pages to explore include Meet Amazing Americans, Jump Back in Time, and Join America at Play.
Thomas J. Elpel's Hollowtop Outdoor Primitive School (www.hollowtop.com). Find out how early humans made fire, tools, and shelter from the natural world around them.

Index

agriculture
  early, 4–5
  today, 30–31
ammonia, 16
aqueducts, 9

barley, 5, 6, 10

cans, 20–21
cash crops, 13
conservation, soil, 26–27
corn, 10, 15
cotton gin, 12
crop failure, 27
crop rotation, 10–11
crossbreeding, 14–15

DDT, 19
drainage systems, 7, 8
drought, 27
Dust Bowl, 26, 27

electricity, 24–25

Fertile Crescent, 4–5
fertilizer, 16–17, 31
food
  genetically modified, 29
  preservation, 20–21, 25
  processed, 30

genetic engineering, 28–29, 30
grain, 5, 6, 7, 13

harrow, 13, 18
herbicides, 29
hybrid, 14, 15, 28

insects, 15, 18
irradiation, 31
irrigation, 7, 8, 9, 25

jars, 20

limestone, 11

machinery, farm, 12–13, 22–23

nutrients, 10, 16, 27

pest control, 18–19, 31
plow, 12
potato, 15, 27

reaper, 13
refrigeration, 20, 21, 24, 25
rice, 6, 7, 8

seeds, 5, 14, 29
shovels, 6, 7
soil conservation, 26–27
steel, 13

threshing machine, 13
tools, 7
topsoil, 8, 10
tractors, 22–23

villages, 6–7, 8

water management, 8–9
weeds, 18, 29
wheat, 4, 5, 6, 10, 13, 14, 15, 26

32